How to create an international production formula for your homemade perfume

Immortalize your perfume now, before it's too late!

Philip Goutell

Lightyears, Inc.
Walden, NY
www.PerfumeProjects.com

Contents

Illustrations

All photography by Philip Goutell

Introduction

Have you ever created a great perfume, one that everybody loved, but you could never make it again because you couldn't remember how you made it? You did it with some essential oils and other aromatics but you can't remember what they were or where you got them. Tragedy.

With the formula in your hands, today you might be making money with that perfume and possibly others that you've made over the years. But – pity! – they are all lost forever because you never took the simple steps that would have made them immortal. *You never recorded the formulas!*

It's so simple

A perfume formula is nothing more than a piece of paper with a few lines on it. Some formulas are worth millions of dollars. Yours could be worth... something.

What you might find surprising is that a perfume formula is more like a recipe in a cookbook than a page out of a physics textbook. A perfume formula is easy to understand. It just shows you how to repeat what you did when you created your perfume.

And there's more

With an accurate formula in your hands you can license others to produce your perfume (they might be really good marketers who can bring in money for you!)

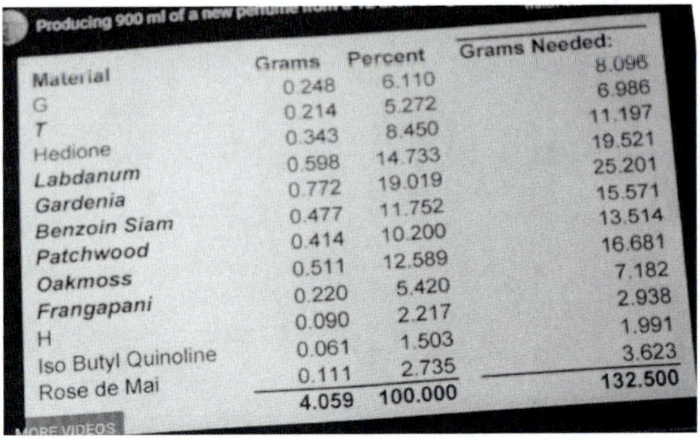

Producing 900 ml of a new p...

Material	Grams	Percent	Grams Needed:
G	0.248	6.110	8.096
T	0.214	5.272	6.986
Hedione	0.343	8.450	11.197
Labdanum	0.598	14.733	19.521
Gardenia	0.772	19.019	25.201
Benzoin Siam	0.477	11.752	15.571
Patchwood	0.414	10.200	13.514
Oakmoss	0.511	12.589	16.681
Frangapani	0.220	5.420	7.182
H	0.090	2.217	2.938
Iso Butyl Quinoline	0.061	1.503	1.991
Rose de Mai	0.111	2.735	3.623
	4.059	100.000	132.500

MORE VIDEOS

This is a perfume formula for one of my fragrances. Notice how several materials have coded designations so I'm not giving it all away. This is a screen shot from a YouTube video.

But that's all in the future. Your first step is to start recording your formulas, to make them immortal! The pages that follow show you how to do it.

If you want to test the commercial power of one or more of the perfumers you've made, *become a formula writer*. As you create a perfume, write down how you created it. What essential oils or other aromatics did you use? How much of each did you use? If you capture this simple information while you are creating, you are only a few steps away from producing a written formula that *could* become a money maker for you.

The pages that follow guide you through the steps that turn the record you made of your perfume's creation into a professional perfume formula.

Chapter 1

First steps toward writing a professional formula for your own perfume

It's all about record keeping and when you're creating a fragrance, record keeping is not difficult. You just have to train yourself to do it. It simply means that, while you are sniffing your way through a new creation, you need to keep a pen and notebook handy. You need to write down all the important stuff. It's not hard. You just have to know

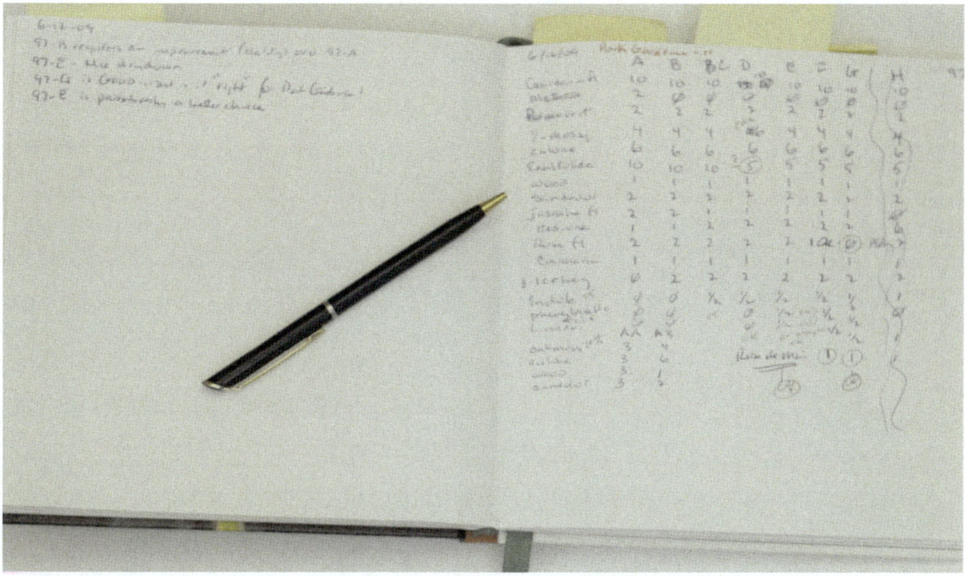

This is a notebook I used while developing several fragrances. Train yourself to write down every trial you make. If you leave out important information you'll be kicking yourself when you hit a winner and are missing some of the details on how it was made.

what to record.

The records you need to keep while creating your perfume are (1) a list of every material used, (2) the amount of each material used, and (3) the identity and *source* of every material you use. With this information you can produce an accurate formula. None of this is complicated. It only requires you to keep your pen and notebook handy while you are creating.

1 List everything that went into your perfume

When you are developing your perfume write down the name of every item you use as you use it. As you adjust your fragrance some items on your list are likely to be crossed off (but just cross them off, don't erase them) and some may be added. Here's how I do it

Each time I try a new material, I write down its full name on my list and, next to that full name, I write a short name that I can use in my daily notes. For example, if I have three aromatics — Cedarwood Virginian Essential Oil, Ambrette Seed Absolute 10% in DPG, and Oakmoss Absolute (IFRA Low Atranol) 10% in DPG — Cedarwood Virginian Essential Oil becomes "Cedarwood," Ambrette Seed Absolute 10% in DPG becomes "Ambrette," and Oakmoss Absolute (IFRA Low Atranol) 10% in DPG becomes "Oakmoss." But always keep a separate record of the *full* description of the material your short name refers to.

These short names will do fine as long as you don't add any conflicting items. For example, if I started to use Ambrette Seed Absolute 1% in DPG while I was using "Abbrette" to denote Ambrette Seed Absolute 10% in DPG, I might revise my short name for it and call it "Ambrette-10" while giving Ambrette Seed Absolute 1% in DPG the short name "Ambrette-1".

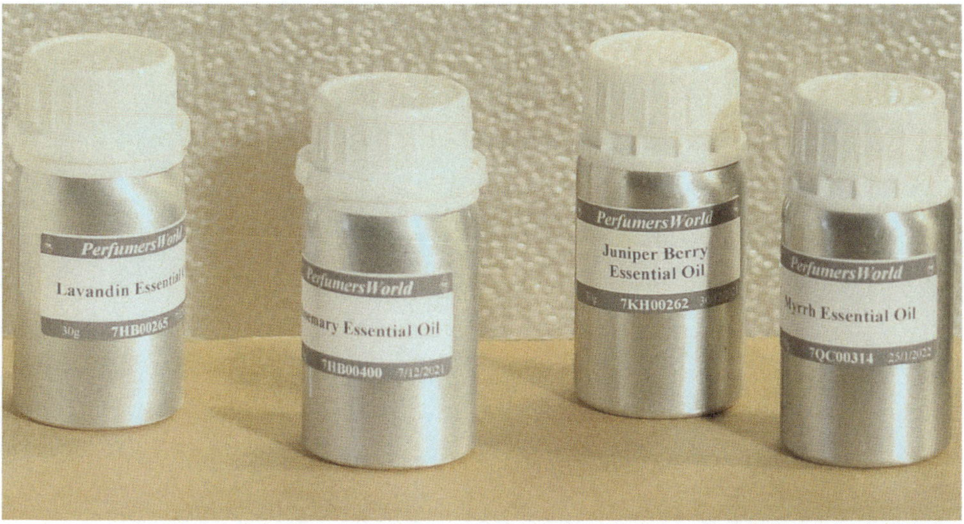

Here are some materials I used for a project. Their full descriptions are: "Lavandin Essential Oil, PerfumersWorld, Code 7HB00256, 7/12/2021," Rosemary Essential Oil, PerfumersWorld, Code 7HB00400, 7/12/2021," Juniper Berry Essential Oil, PerfumersWorld, Code 7KH00262, 7/12/2021," and "Myrrh Essential Oil, PerfumersWorld, Code 7QC00314, 7/12/2021." These descriptions identify the product, the source (PerfumersWorld), the vendor's code, and the date these small cans were filled by the vendor. In my daily notes I used the short names "Lavandin," "Rosemary," "Juniper," and "Myrrh."

Note:

DPG is the common abbreviation for Dipropylene Glycol, a nearly odorless solvent used to reduce the strength of odorants. "Ambrette Seed Absolute 10% in DPG" would indicate that the oil is 10% Ambrette Seed Absolute and 90% DPG. Using Ambrette Seed Absolute at 10% would give you quite different results than using Ambrette Seed Absolute at full strength.

"Oakmoss Absolute (IFRA Low Atranol)" indicates that this Oakmoss has been processed to IFRA (The International Fragrance Association) standards to reduce irritants.

Short names are useful because they save time when taking notes. When I'm working on a fragrance I'm always a bit rushed, eager to discover the effect of new materials and new combinations of materials. Yet I want to maintain an accurate record of each new material and each new combination I try. If I had to write down the full name each time I used an ingredient I would probably get sloppy and in time my notes would become less accurate. Using short names in my notes is a way to maintain their accuracy.

2 Record the amount of each material you used

Here comes the first technical issue. You must record the *amount* of each material used but how can you do it? How can you make accurate measurements? A global fragrance company may measure ingredients by weight. For them in making a trial some ingredients will be used in a very small weights but *they* have the equipment needed to measure very small weights. *Their* equipment costs tens of thousands of dollars. What *they* use to make measurements is of no help to you.

If you can't accurately weigh the small amounts of material you are using could you possibly measure them by volume? Here again you run into the problem of not having the equipment needed to accurately measure fractions of a fluid ounce or milliliter. But you *can do* exactly what many perfume developers do. You can take your measurements in *drops* squeezed out of dropper bottles.

Working with dropper bottles

For creators of perfume who do not have access to a sophisticated laboratory (and that includes most of us), developing a new fragrance by counting out drops from dropper bottles is a standard technique. Dropper bottles allow you to work with reasonable accuracy with small quantities of material. This is important because some materials you might want to use – certain essential oils for example – can be quite expensive. You will want to limit the amounts used in your experiments, your "trials" as they are called. But to get the most accurate measurements possible with dropper bottles there are guidelines to follow.

Rule # 1 — All your bottles and droppers should be the same size and come from the same source. Some of the materials you purchase may come in small bottles without droppers, some may come in small dropper bottles, others may come in large dropper bottles. Since your goal is for all drops to be as equal in size as possible, your droppers and dropper bottles themselves should be uniform in size and newness. You can achieve

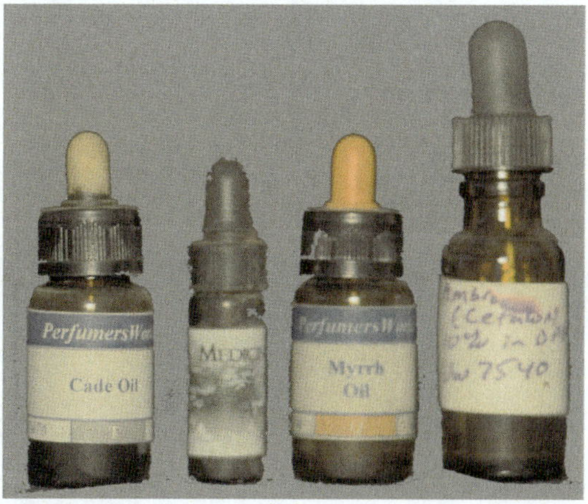

Notice how these bottles differ in size. Drops from their droppers will also differ in size. Notice also that the dropper on the left is discolored from age. Although it is the same size as the dropper second from right (Myrrh) its drops may be irregular in size. Your goal when using dropper bottles is for all the drops to be as close to the same size as possible.

One way to insure that all your drops are close to the same size is to put your aroma materials in new dropper bottles with new droppers.

this by purchasing a quantity of bottles with matching eyedroppers and then transferring your materials to these bottles. Then not only will the glass tubes of the droppers be uniform but the rubber bulbs will all have close to the same resistance when squeezed.

HINT: be sure to label each dropper bottle carefully with what you have put in it.

Rule # 2 — As your goal is to make the drops from your dropper bottles as uniform in size as possible, make it a habit to hold your droppers in the same position when squeezing out drops. Some say the droppers should always be held straight up. I have found that for some thin, runny materials this allows unwanted drops to drip from your dropper throwing off your count. I personally hold my droppers at about a 45 degree angle. This helps

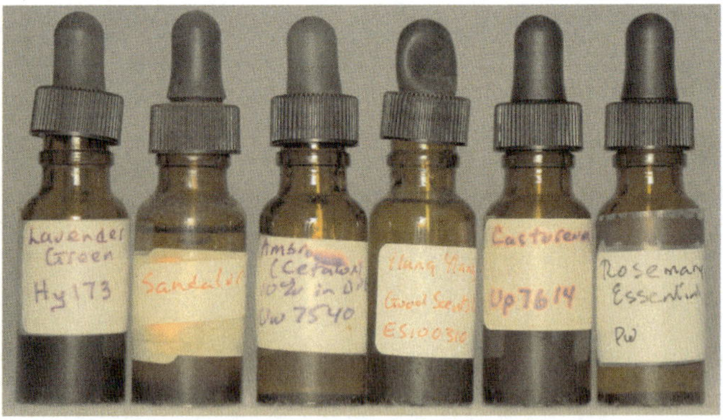

When you transfer materials to new bottles be sure to label them. Notice here the bottle third from the right. The bulb of the dropper is distorted and the dropper should be replaced with a new one. The bottle itself is fine.

prevent those extra drops from escaping on their own.

I was taught that half drops can be achieved by slowly releasing a bit of material onto a toothpick and allowing less than a full drop to fall into my container. Another trick for which accuracy can't be guaranteed is, when you want the smallest amount possible of a material, just wet the tip of a toothpick and then stir it into your mixture. This might be considered 1/1000th of a drop and, while the technique has some merit, it doesn't lend itself to careful record keeping. However, if you ever get a chance to inspect a formula by a professional perfumer, you may see certain materials being used in seemingly microscopic amounts. Personally I try to avoid working with anything smaller than half a drop unless I'm creating a small trial for my own pleasure.

Rule # 3 — Record the source and identity of every material you use. The issue is *continuity*. To *make* more of the same you have to be able to *buy* more of the same. When shopping for materials always ask yourself, "Will I be able to get more of it?"

You are most likely to be able to get more of the materials you are using if you buy them from established sources that continue to offer them year after year. For synthetic materials look for CAS numbers. They are precise identifiers.

CAS numbers

The *Chemical Abstract Service,* founded in 1907 and

now known as the *CAS Registry*, assigns numbers to chemical compounds "that have been described in literature." For you the significance of CAS numbers is that they are a way to communicate the identity of aroma materials to a third party.

Essential oils generally will *not* have CAS numbers since by their nature they are imprecise blendings of multiple chemicals and can differ greatly from one batch to the next depending on how they were processed, where the material was harvested, and even the season and year in which the material was harvested.

Vendors that serve major fragrance houses try to iron out inconsistencies between batches of essential oils produced in different years or seasons by carefully blending them.

Your best hope for consistency is to buy from a vendor that serves commercial fragrance houses.

For materials without a CAS number, your records should identity them by vendor's name, the aroma material's name, and the vendors stock or catalog number.

A particularly important issue with continuity occurs when you have a third party producing your perfume. You don't have to be a big company to have your perfume produced by third party professionals. Many of my fragrances have, after I signed off on a small sample, been produced by for me by a third party, using one of my formulas. But to produce your fragrances successfully, along with your formula you have to give the production team the *exact* identity of the materials you have used so they can find the same.

Problems to look out for

I have a variety of essential oils on hand that I have been given as samples or purchased out of curiosity to sample a particular aroma. None of them will ever be used in a perfume I create. Why? Simply because I can't be sure exactly what they are and I can't be sure I can get more of the exact same. When you buy the aroma materials you are going to work with, give some thought to their source. Your best bet is to buy off a well respected website, one that has been around for a number of years, one that has

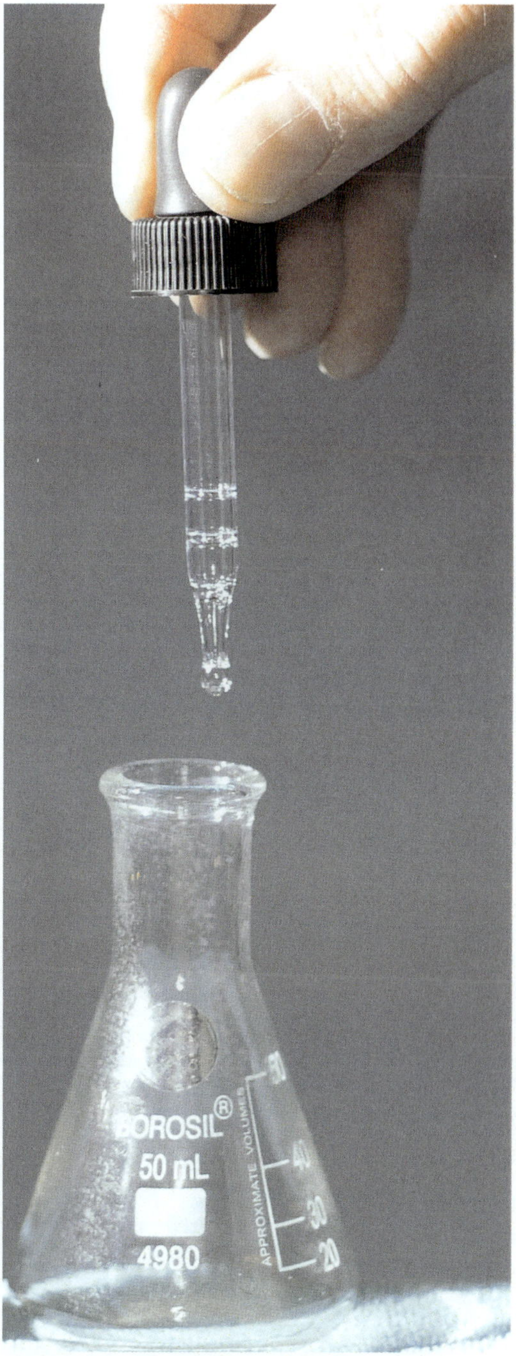

Holding the dropper vertical is a recommended practice. In my opinion this does not work well with thin, runny materials.

I like to hold my droppers at an angle as, from my experience, thin, runny materials are less likely to add unwanted drops. Whichever way you choose to hold your droppers, be sure to hold them in the same position for all your drops from all your bottles.

been recommended by others, and one that gives you solid information about their products.

Avoid oils sold in health food stores unless they are from a very well known brand. In spite of what you read on the label you can't really be sure of what you are getting, especially when you are buying a blended oil. How much ingredient "A" does it contain? How much of ingredient "B"? Will the ratio be consistent?

Beware of essential oils advertised as authentic but are being sold at too low a price. You may get a little of what is advertised on the label but you might get more of other things along with it and you can't be sure of what they are.

Chapter 2

Writing your formula in drops prior to weighing it

Written in drops, your formula could look like this:

MATERIAL	DROPS
Clove Bud Oil	7
Ambroxan	1
Lyral	2
Benzyl Salicylate	8
Cade Oil	1
Sandalor	12
Ethylene Brassylate	11
Total =	**42 drops**

As written, this formula will produce just 42 drops. Not very much. To produce a larger quantity of this fragrance we could multiply the number of drops of each material by some number, large or small, depending on how much of our perfume oil we wanted to produce. Say we wanted to produce 4,200 drops of our fragrance. Our formula would now look like this:

MATERIAL	DROPS x 100
Clove Bud Oil	700 drops
Ambroxan	100 drops
Lyral	200 drops
Benzyl Salicylate	800 drops

Cade Oil	100 drops
Sandalor	1200 drops
Ethylene Brassylate	1100 drops
Total =	**4200 drops**

But look at the tedious work and the intense concentration required to produce this larger quantity – which really isn't so large at all. While we only have to count out 100 drops of Ambroxan and 100 drops of Cade Oil, we will have to count out 1200 drops of Sandalor and 1100 drops of Ethylene Brassylate. It's highly probable that you could loose track of your count somewhere around the 300 drop mark. Then you would have to throw out the drops you have already counted (since you can't be sure of their quantity) and start all over again – and likely you will encounter the same problem again.

Working with percentages

If we knew the percentage of the weight of the finished fragrance each material represented, we would not have to count drops. We would take the weight of the fragrance oil we wanted to produce and then combine the materials in the formula by the percentage of the weight of the formula each material represented. In short, we would have a formula we could use to produce *any* amount of fragrance that we wanted using *any* measurement of weight requested (kilos, pounds, tons, etc.).

For example, if we knew that our fragrance was a mixture of A,B, and C and, by weight, A was 20 percent, B was 37 percent, and C was 43 percent, then to make 10 pounds of the fragrance oil we would simply combine 2 pounds of A, 3.7 pounds of B, and 4.3 pounds of C to give us our 10 pounds. (2 + 3.7 + 4.3 = 10)

Moreover, once this formula was written in percentages of weight it wouldn't matter what unit of measurement we worked with – grams, kilos, ounces, pounds, or even tons. The percentage of each material by weight will be the same.

Converting a formula in drops to a formula in percentages of weight

To convert our formula in drops to a formula in percentages of weight we start by weighing our drops. The weight of any one drop will be too small to measure accurately but if we multiply the number of drops of each material by a number that will allow us an accurate measurement of the *smallest* material, weighing becomes practical.

Multiplying the number of drops in the formula above by 100 would make it practical to measure the weights of Ambroxan and Cade Oil but counting out 1100 drops of Ethylene Brassylate and 1200 drops of Sandalor would be asking for trouble. However if, rather than 100, we multiply each number of drops by 20, 20 drops of Ambrox and and Cade Oil will give us enough weight for accuracy while counting out 220 drops of Ethylene Brassylate and 240 drops of Sandalor are quantities small enough to allow us to keep our focus when counting. Now our formula will look like this and be ready to be weighed using equipment we can afford:

MATERIAL	DROPS x 20
Clove Bud Oil	140 drops
Ambroxan	20 drops
Lyral	40 drops
Benzyl Salicylate	160 drops
Cade Oil	20 drops
Sandalor	240 drops
Ethylene Brassylate	220 drops
Total =	**840 drops**

To weigh these drops we will use an electronic balance.

Chapter 3
Introducing the electronic balance

An electronic balance is a tool you can use to weigh the aroma materials used in your perfume. It is, in essence, a scale and not so different than a bathroom scale that weighs people or a kitchen scale that weighs food. Weights of "mass" — the objects being weighed — are displayed digitally. If you are a perfume creator, you may need at least two electronic balances, each with a different

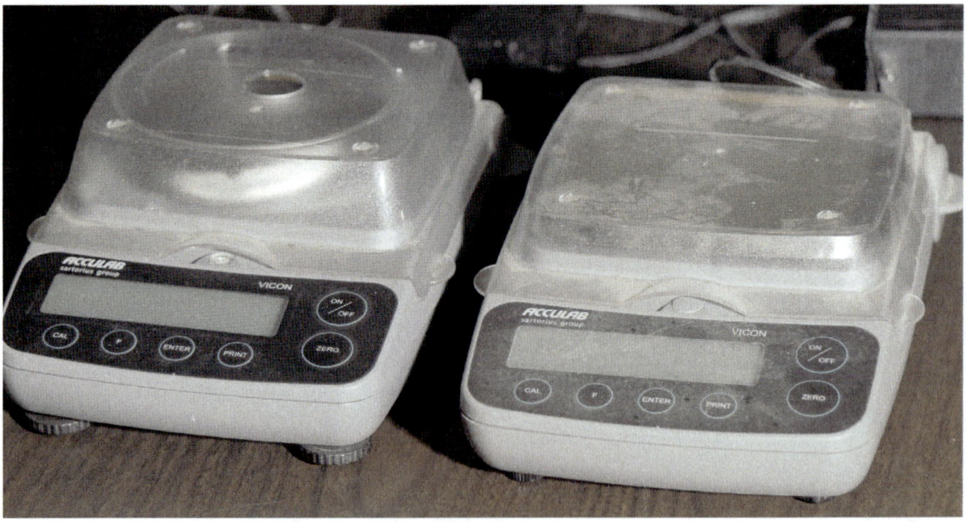

Here are the two electronic balances I use when developing a perfume. The one on the left has 3-digit (0.001) readability and a weighing capacity of 120 grams. The 120 gram capacity is too small for weighing the elements of production batches so I have a second electronic balance (seen on the right) which has a weighing capacity of 3100 grams – but only 1-digit readability (0.1). This model lacks a leveling bubble but an inexpensive leveling bubble from a camera store can do the job.

weighing capacity and different degree of precision.

Purchasing an electronic balance

Cost

As I write this in 2022 you should be able to purchase a good quality electronic balance with all the features you'll need for between $200 and $300. As noted above, you will probably need two of them.

Readability

The first electronic balance you will need is one that will take measurements in grams down to 1/1000th of a gram (0.001 gram) This is referred to as 3-digit (or 3-decimal place) readability. When zeroed the digital display will show 0.000.

Weighing capacity

A balance with 3-digit readability is likely to have a limited weighing capacity. My 3-digit balance has an upper range of just 120 grams. This is not enough for me when I'm producing a perfume. Therefore I also use a balance with a 3100 gram capacity but this balance has only 1-digit readability meaning it will only read down to 1/10th of a gram (0.1 gram) so I can't use it for weighing small quantities. Search as you will, you're not likely to find a single balance that has a large weighing capacity and will weigh down to 1/1000 of a gram.

A/C Adapter

Even when your electronic balance can be powered by batteries, an A/C adapter is almost an essential. No fuss with weak batteries right in the middle of weighing out 100 drops or other inconveniences. Just plug in the adapter, turn on your balance and you're ready to go for as long as your work takes.

Electronic balances in the $200 to $300 price range typically can be powered by batteries. This makes them suitable for field use, away from electric outlets. Most also come with A/C adapters and an A/C adapter is essential (or at least very useful) for laboratory use. For me this is a "must have."

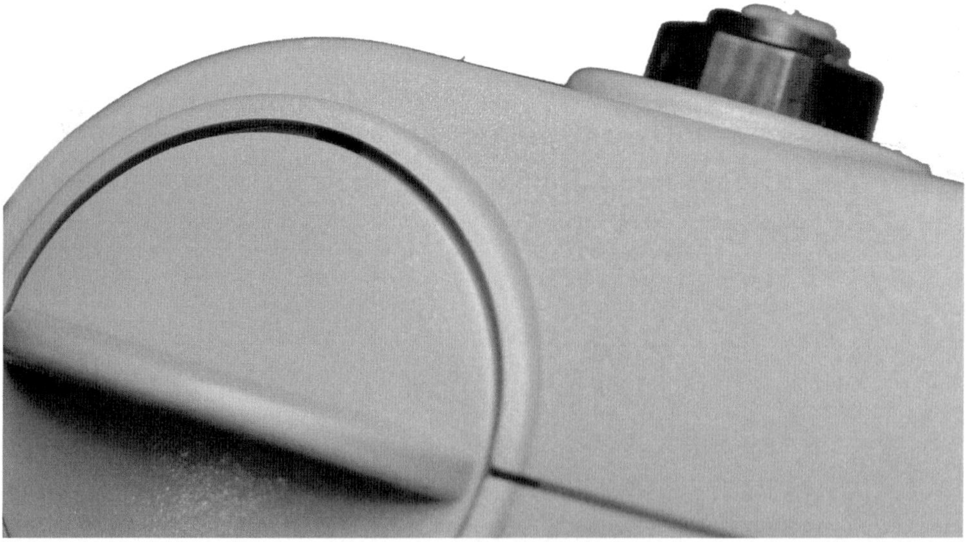

To be leveled (which is essential in weighing), your electronic balance requires four adjustable feet; one of which is shown here. The circle to the left is the door to the compartment where the 100 gram calibration weight is stored.

Adjustable feet and leveling bubble

To get accurate results, your balance has to be leveled. Adjustable feet on the balance are required to level it. When the bubble is centered, your balance is level.

You need to level your balance before weighing anything but some balances are not equipped with leveling bubbles. In this case you can get a leveling bubble (for a camera) from a camera store. Place it in the center of the weighing pan of your balance and adjust the feet of your balance until the bubble is centered.

This is the 100 gram calibration weight for my 3100 gram capacity balance. During the calibration process it is placed in the center of the weighing pan. After use it is stored in a compartment inside the rear of the balance.

Weighing pan

The balance should have a pan on which you will place the objects you are weighing (which will usually be glass containers of some sort.)

Calibration

Your balance will commonly be provided with one or more calibration weights. These are standard weights used when calibrating the balance, an important step

When you move your balance to a new location, or if you haven't used it for a while, it should be recalibrated. This involves placing a calibration weight in the center of the pan of the balance and following the calibration instructions in your user's manual. To the right of the picture you can see the removable pocket in which the calibration weight is stored.

before use. The user's manual for the balance will explain the calibration process.

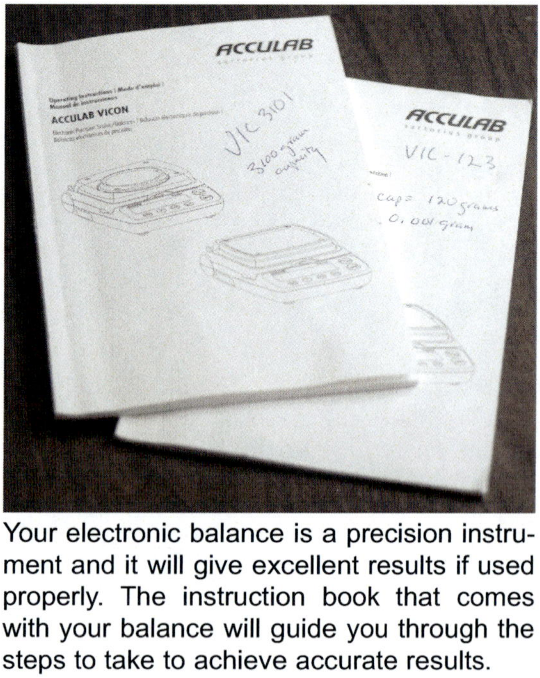

Your electronic balance is a precision instrument and it will give excellent results if used properly. The instruction book that comes with your balance will guide you through the steps to take to achieve accurate results.

User's Manual

You should get a user's manual with your balance. This is important and it will explain in greater detail the setup, calibration process and use of your balance. It is also likely to introduce you to capabilities not men-

The front panel of this electronic balance looks quite simple but there's a lot going on. To the right of the digital readout is the on/off button and below it the button to zero the balance. The smaller buttons from left to right are for calibration (CAL), function (F), enter (ENTER) and print (PRINT). You will make frequent use of the calibration button but perhaps only one-time use of the function button, to select the units of weight you want displaned – grams, ounces, or one of a number of other choices. The ENTER and PRINT buttons would be used only if the balance was connected to an Acculab technical printer.

tioned here. You may never need them but it's nice to know they exist and how you could access them.

How to use an electronic balance

Read the instructions that come with your balance. Set up your balance on a flat, solid surface. Calibrate your bal-

Once your balance is set up and turned on, it is zeroed by pressing the large button at the right. The display will show all zeros and, once it is stable, the small "balance" icon at the left will appear. After about three seconds the readout will begin to drift and the icon will disappear.

The display is showing a weight of 37.232 grams for the flask on the weighing pan of the balance. The "scales" icon is not showing, suggesting that either the display has not yet stabilized or it has already stabilized and then begun to drift.

ance following the instructions in your manual. Be sure that your balance is set to measure the correct weight units (grams, ounces, or something else.)

Start your weighing by zeroing the balance. Place the object to be weighed on the pan of the balance. Wait about three seconds while the display stabilizes. (There may be a symbol in the display that will appear when the display is stable.) Read and record the weight shown in the display. That's all there is to it.

Best practices

The quality of the perfume you produce will depend on the quality of your measurements, both in developing your formula and in producing your finished perfume. Here are some points to be aware of.

Location — Set up your balance on a solid, flat topped table or bench. Avoid setting up your balance on any furniture with a wobble. A heavy table or workbench is ideal.

Environment — Avoid areas of high foot traffic Avoid vibrations from boom boxes, radios, stereos, etc. Keep your balance out of drafts from windows, heating systems, doors, or air conditioners. A draft can throw off your measurements.

Warm up — Allow your balance to warm up before use. Both your balance and the materials you are weighing should be at room temperature. Changes in temperature can affect accuracy.

Contamination — Once you have weighed the container you are using to weight your materials, avoid setting it down on a counter where the bottom of the container may pick up a small amount of unseen debris. For super accuracy (when weighing with a device that goes beyond the 3-digit limit) avoid touching the weighing container with your hands. Use tongs or gloves or a piece of paper wrapped around the measuring container.

Clean up — Clean your balance as soon as you have finished using it.

Moisture — Avoid using your balance in damp areas. Moisture, such as small amounts of condensation, can add small, unwanted weight.

Recalibrate — Recalibrate your balance if it is moved from one location to another.

Room temperature — Early mornings can be cool or hot. Heating and cooling systems may be turned off overnight. By afternoon your laboratory temperature will have adjusted. This is a good time to do your weighing.

Use the center of your weighing pan — Place your weighing container on the center of the pan. If the container is far off center, errors creep in.

More than you really need to know

An electric balance measures the force applied to the pan. That force is converted to an electrical signal and displays on a digital display.

An electronic balance is subject to gravitational pull. Readings can change depending on latitude and altitude which is why repeated calibration is essential.

If your balance fails to stabilize, moisture could be the culprit.

All measurements have some degree of uncertainty. Measuring several times and averaging the results is one way to limit uncertainty.

Analytical balances that measure to four places (0.0001 grams) generally cost from about $1500 to $40,000 or more and require regular professional calibration.

You cannot control all sources of error. Multiple trials and averaging can help control errors, to a degree. The percentage of error you can expect can be calculated mathematically but I can't explain how this is done.

Even an accurate balance will drift after a number of seconds. Note the weight on the display and write it down before your display drifts.

Plastic and glass weighing containers can hold an electric current that can cause inaccuracies in weighing.

Chapter 4

Weighing drops and calculating percentages

Now, using an electronic balance, we can weigh our "drops times 20" formula and then calculate the percentage of the weight that each material in our formula represents.

First we weigh our drops:

MATERIAL	DROPS	WEIGHT IN GRAMS
Clove Bud Oil	140 drops	55.563
Ambroxan	20 drops	16.410
Lyral	40 drops	1.581
Benzyl Salicylate	160 drops	22.200
Cade Oil	20 drops	33.370
Sandalor	240 drops	40.115
Ethylene Brassylate	220 drops	87.313
Total =	**840 drops**	**256.562 grams**

Then we calculate what percentage of the formula those weights represent:

MATERIAL	GRAMS	PERCENT OF FORMULA
Clove Bud Oil	55.563 / 256.562 x 100	= 1.657
Ambroxan	16.410 / 256.562 x 100	= 6.396
Lyral	1.581 / 256.562 x 100	= 0.616

Benzyl Salicyl	22.200	/ 256.562	x 100 =	8.653
Cade Oil	33.370	/ 256.562	x 100 =	3.001
Sandalor	40.115	/ 256.562	x 100 =	15.636
Ethylene Brassylate	87.313	/ 256.562	x 100 =	34.032
Total =	**256.562 grams**			**99.991**

A quick review of what we've just done:

First we weighed the drops of each material and recorded the result in grams. Then, to calculate the percentage of the formula (by weight) of each material represented, we divided its weight by the total weight of all the materials (256.562 grams) and then multiplied the (*decimal*) result by 100 to give us the *percentage* of the formula (by weight) that material represented.

The total of all the percentages should come to 100 but numbers had been rounded off, both in weighing and in multiplying. Coming up with a total of 99.991 percent is a good result and we are only slightly off the "100" mark.

And now we can produce any weight of our formula we want

For example, suppose we wanted to produce 15 kilos of our formula. The math would look like this:

MATERIAL	PERCENT	WEIGHT NEEDED
Clove Bud Oil	21.657% of 15 =	3.25 kilos
Ambroxan	6.396% of 15 =	0.96 kilos
Lyral	0.616% of 15 =	0.09 kilos
Benzyl Salicylate	8.653% of 15 =	1.30 kilos
Cade Oil	13.001% of 15 =	1.95 kilos
Sandalor	15.636% of 15 =	2.35 kilos

Ethylene Brassylate	34.032% of 15 =	5.10 kilos
Total =	**99.991%**	**15.00 kilos**

Quick review of what we've just done

To produce a total of 15 kilos of our formula we calculated the weight each of our materials contributed to the 15 kilo total using the necessary percentage of each material we had already calculated.

The importance of what we have done

By translating our original "drops" formula into a formula of percentages of weight we can now calculate the weight of each material needed to produce this perfume oil in any amount and in any units we choose. ***The formula is now universal.*** Anyone with a copy of this formula and a knowledge of the exact materials required can recreate our fragrance oil simply by combining the materials.

Produce, wait 30 days, and then test it

With your new "by percents" formula in hand I suggest you test its accuracy by making a small batch of your perfume oil, perhaps half a kilo (500 ml). Then, as always when you mix aroma materials, wait 30 days to give them time to blend. Shake them up from time to time while they are blending. At the end of 30 days (or a bit more) sample what you have made by comparing its scent to that of a small sample made from your original "drops" formula. How closely do the scents match?

Ideally you will get a good match. If so, you're on your way to making as much of your perfume oil as you want. But it is also possible that this new batch may be slightly off in one or more notes. Now, with the aid of your nose, you can fix *both* the batch of perfume oil you just made *and* your formula.

You be the judge

Is one note off? Not strong enough or too strong? To fix a note that is not strong enough, one that gets buried in the background scent, add more of it to your finished perfume oil. You might start by adding 5 percent of what is called for in your formula. Then, if needed, add another 5 percent, each time recording the weight of what you added. When this procedure brings the scent up to what was wanted, revise your formula to account for this change. Here is an example of how to do it.

Referring to the sample formula above to make 15 kilos of fragrance oil, say the Ambroxan was too weak and we added 5 percent more (0.14 kilos) and then another 5 percent (0.14 kilos). Now, instead of 0.96 kilos of Ambroxan we have used a total of 1.24 kilos, so the weights of the materials used and our new percentages look like this:

MATERIAL	KILOS	NEW PERCENTAGE OF FORMULA*
Clove Bud Oil	3.25 / 15.28 x 100 =	21.27 %
Ambroxan	1.24 / 15.28 x 100 =	8.12 %
Lyral	0.09 / 15.28 x 100 =	0.59 %
Benzyl Salicylate	1.30 / 15.28 x 100 =	8.51 %

Cade Oil	1.95 / 15.28 x 100 =	12.76 %
Sandalor	2.35 / 15.28 x 100 =	15.38 %
Ethylene Brassylate	5.10 / 15.28 x 100 =	33.38 %
Total =	**15.28 kilos**	**100.01%**

**Numbers have been rounded to 2 decimal places.*

If, on the other hand, one material was too strong, since we could not subtract it from what was already made, we would have to raise the level of all the other materials, little by little, until the correct balance was achieved. Our work could look like this if, referring to our formula again, we found that the Clove Bud Oil was too strong and by raising the levels of all the other ingredients by 5 percent, harmony could be achieved. Then our new formula would look like this:

MATERIAL	**KILOS**	**NEW PERCENTAGE OF THE FORMULA**
Clove Bud Oil	3.25 / 15.60 x 100 =	20.83 %
Ambroxan	1.01 / 15.60 x 100 =	6.47 %
Lyral	0.09 / 15.60 x 100 =	0.58 %
Benzyl Salicylate	1.37 / 15.60 x 100 =	8.78 %
Cade Oil	2.05 / 15.60 x 100 =	13.14 %
Sandalor	2.47 / 15.60 x 100 =	15.83 %
Ethylene Brassylate	5.36 / 15.60 x 100 =	34.36 %
Total =	**15.60 kilos**	**99.99%**

**Numbers have been rounded to 2 decimal places.*

When adjusting the scent you produced remember to record the weight of each material you added so that you can recalculate the percentages of each material used and create a new, more accurate formula. When you have done this, the next time you produce a batch of your fragrance oil the results should be spot on.

Chapter 5
Sometimes complications arise and drops present a problem

You have seen now how a formula originally created in drops can be weighed out in grams and then converted to percentages of weight. This eliminates the necessity of counting out huge numbers of drops as when we weigh, we are only dealing with, at most, a few hundred drops of any particular material. But look at this formula. Notice the problem it presents:

Formula 6

Material A = 60 drops
Material B = 120 drops
Material C = 45 drops
Material D = 3 drops
Material E = 70 drops
Material F = 1 drop
Total = 299 drops

To get a large enough quantity of Material F to weigh accurately, we have to multiply all the quantities by some number. Twenty would be the smallest multiplier I would use as it brings Material F up to 20 drops, enough for accurate weighing. But look now at what happens to Material B: 120 x 20 = 2400 drops. And look what happens to Material E: 70 x 20 = 1400 drops. And Material A: 60 x 20

= 1200 drops. I personally don't trust my hand and my eyes measuring out 1200 drops, 1400 drops, and certainly not 2400 drops. So here I'm going to use a different strategy.

I would not recommend this alternate strategy in most situations but counting out a few thousand drops has no appeal to me. Thus I'm going to make my production formula using a different method. However, I have two warnings for you before *you* use this alternative. First, the drops measured out in your original formula must be as uniform as possible. This means you took care to use the same size bottles and droppers for all your materials and you always held the dropper in the same position when squeezing out drops. Then, for this alternative method, you must have a good enough nose to be able to judge whether this method has given you the same scent as your original "drops" formula and, if it has not, you must have a good enough nose to make any needed adjustments.

So here we go.

Rather than take our percentages from the weight of grams, we'll take our percentages directly from the formula in drops:

Table 2

Material A: 60 drops	60/299 x 100 =	20.07 percent
Material B: 120 drops	120/299 x 100 =	40.13 percent
Material C: 45 drops	45/299 x 100 =	15.05 percent
Material D: 3 drops	3/299 x 100 =	1.00 percent
Material E: 70 drops	70/299 x 100 =	23.41 percent
Material F: 1 drop	1/299 x 100 =	0.33 percent
Total drops = 299	**Total percents = 99.66**	

Now we are ready for production. Since it is just the perfume oil we are producing and, in our finished perfume it may be as little as 7 percent of the mixture (alcohol and water supplying the other 93 percent), we will use our production formula to make just 800 milliliters (0.8 liter) of our perfume oil. Remember, we are now working with *volumes* of liquid rather than weights.

So we now calculate:

FORMULA # 2

Percentage of formula by volume		**Volume required to make 800 ml of fragrance oil**
20.07 percent of Material A =	.2007 x 800 =	160.56 ml
40.13 percent of Material B =	.4013 x 800 =	321.04 ml
15.05 percent of Material C =	.1505 x 800 =	120.40 ml
1.00 percent of Material D =	.01 x 800 =	8.00 ml
23.41 percent of Material E =	.2341 x 800 =	187.28 ml
0.33 percent of Material F =	.0033 x 800 =	2.64 ml
	Total =	**799.92 ml**

Measurement tools needed

To measure by *weight* we used an *electronic balance*. To measure by *volume* we will use *graduated cylinders and a kitchen measuring cup*. This equipment is far less expensive than a single electronic balance. Here are the items you now need to add to your perfume laboratory. They all fall under the category of "glassware."

Essential glassware

4-cup Pyrex glass measuring cup. This is something

you can find in your local supermarket. I have several of them and they get a lot of use. You want one that will measure in both fluid ounces (up to 64 fluid ounces) and milliliters, up to 1000 milliliters (1 liter).

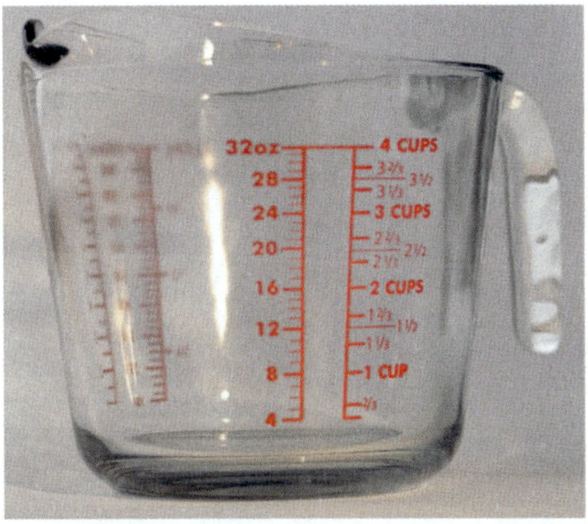

4-cup glass measuring cup showing graduations in fluid ounces

Small measuring column. A graduated measuring column that can measure up to 10 milliliters.

Medium measuring column. A graduated measuring column that can measure up to 25 milliliters.

Large measuring column. A graduated measuring column that can measure up to 50 milliliters or more.

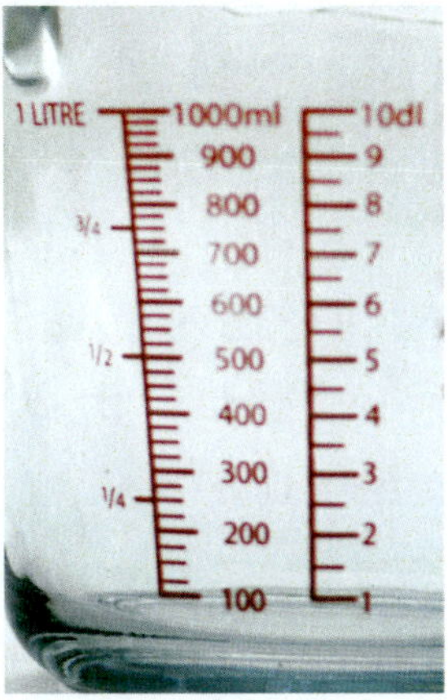

Reverse side of measuring cup showing graduations in milliliters

These measuring columns can often be obtained as a set along with other laboratory glassware. A set will often include calibrated beakers of various sizes and these can be useful as containers when you are developing your perfume.

Now let's make use of this measuring equipment to prepare the 800 ml batch of the formula above. Because this is not a "real" perfume, I won't use "real" materials but you can follow these instructions using colored water.

Here is how you assemble your "fragrance"

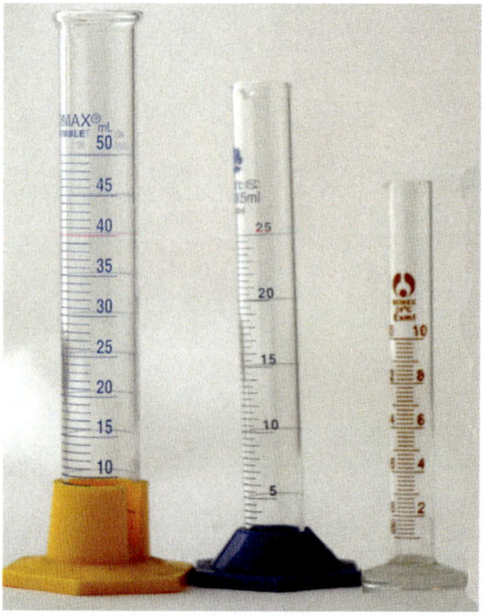

3 measuring cylinders

For Material A you need 160.56 ml so fill you large measuring column to the 50 ml mark three times, each time dumping the contents into your 4-cup measuring cup. This will give you 150 ml. To get your final 10.56 ml use your medium measuring column. Fill it to the 10 ml mark and then add just a little more for the extra half you need. Pour this into your 4-cup measuring cup.

For Material B you need 321.04 ml so fill your large measuring cup to the 50 ml mark six times, each time emptying the contents into your 4-cup measuring cup. Then use your medium measuring cup and fill it to the 21 ml mark. Add the contents to your 4-cup measuring cup.

For Material C you need 120.4 ml so fill your large measuring column to the 50 ml mark twice, each time dumping the contents into the 4-cup measuring cup. Then use your medium measuring column to measure out the additional 20 ml and dump this into the 4-cup measuring cup.

For Material D you need 8 ml so use your small measuring column. Fill it to the 8 ml mark and dump the contents into your 4-cup measuring cup.

For Material E you need 187.28 ml so use your large

measuring column. Fill it to the 50 ml mark three times, each time dumping the contents into your 4-cup measuring cup. Then, again using your large measuring column, fill it to the 37 ml mark and then dump the contents into your 4-cup measuring cup.

For Material F you need 2.64 ml so use your small measuring column filling it slightly above where the 2.6 ml mark would be (if there was one!) Dump the contents into your 4-cup measuring cup.

Now look at the graduations on your 4-cup measuring cup. It should be filled to close to the 798.6 milliliter mark.

If this was real

If you have been mixing real fragrance ingredients instead of colored water, at this point you would have slightly less than 800 milliliters of your fragrance due to rounding errors and small errors in measuring. You would now pour your finished fragrance oil into a bottle, screw on a cap, shake it a bit and let it age for 30 days or more periodically giving it a few shakes.

> NOTE: For this I use 1-quart (32 ounce) amber Boston round bottles that come with matching screw-on tops.

But you're not finished yet

The final challenge is to have the patience to wait out your 30 days or a bit longer and then sample what you have made. Here's where you'll need a good nose and some of your original fragrance made by drops. Does the perfume oil in your quart bottle have the same scent as your original composition? Is it, as far as you can tell, a very good match? Or will you need to make some adjustments?

After 30 days Making corrections

Thirty days after you produced a batch of your perfume oil based on calculations of the volume of your drops it is time to test your results.

You test using your nose. Does the scent of the batch you just made match the scent of your original "drops" formula? Try them both on smelling strips. Are they a good match? If they are not, you'll need to make some corrections.

The good news is that the most likely source of trouble will be materials used in the smallest quantity but before you begin to tinker, be sure that the aroma materials used in your production batch were identical to the materials used in your original batch: same name, same source and, if you have it, same CAS number.

Hopefully your nose is up to sniffing out what, if anything, is not quite right with the fragrance you just produced. If, in fact, it gave you what you wanted or very close to it, stop! There is no sense in messing with success. On the other hand, if what you made is not quite perfect even though you followed your own formula carefully, you'll want to make adjustments. But the questions is, "what isn't quite right?" Which note or notes are not coming through? Or which note or notes are coming on too strong? You are trying to achieve the same balance you had in your original "drops" batch.

At this point it is important to know the scents of the individual materials that went into your perfume and what each one was supposed to do. With this knowledge you can spot the notes that are out of sync and correct them. You may have several notes out of adjustment but let's take the case of a single note that came out too weak. This means you need more of it in your formula and here's a way to achieve that.

NOTE: These adjustments are similar to the adjustments made in Chapter 4 but are measured in volume — milliliters — rather than in grams or ounces.

Suppose after producing 800 ml (or 799.92 ml of Formula # 2 we found Material D wasn't coming through properly and needed to be increased. Five percent of Material D would be 0.4 ml so we would add that to the production batch. And say we needed to add another 0.4 ml to that batch to achieve the scent we were looking for.

Now, rather than having used 8 ml of Material D we have used 8.8 ml. We have corrected the scent of the production batch by adding this extra 0.8 ml of Material D but now we have to correct the formula itself. Here is how we do this:

FORMULA # 3

Material	**Volume used to make 800 ml of fragrance oil**		**New Formula**
Material A =	160.56 ml.	160.56 / 800.72 x 100 =	20.05 %
Material B =	321.04 ml.	221.04 / 800.72 x 100 =	40.09 %
Material C =	120.40 ml.	120.40 / 800.72 x 100 =	15.04 %
Material D =	8.80 ml.	8.80 / 800.72 x 100 =	1.10 %
Material E =	187.28 ml.	187.28 / 800.72 x 100 =	23.39 %
Material F =	2.64 ml.	2.64 / 800.72 x 100 =	0.33 %
			Total = 100 %

If, rather than a note being missing, a note or notes are too prominent, the adjustment requires that you raise the level of the non-offending notes. This can be a bit more complicated, especially if you have used a number of different materials, but it is the only solution. Again you'll need to record exactly what has been added and, when you are satisfied with the scent you have achieved, rework the percentages in your formula to reflect the changes.

Chapter 6
Finishing your perfume by adding alcohol

When you finish your perfume by adding alcohol you find yourself working with volumes. If, for example, you want to produce 40 gallons of perfume at 10 percent fragrance oil to 90 percent alcohol (a common ratio), your volume of oil will simply be four gallons, ten percent of forty.

If you created your formula by percentages of volume, you just do a liters to gallons conversion and you're set to go.

NOTE:
1 liter = .2642 gallons
1 gallon = 3.7843 liters

But what if you followed my recommendation and produced your fragrance by weight? With your formula you can produce any weight of your fragrance oil you need but what weight will give you four gallons? The calculation is not difficult.

First, you need to know the weight of a volume of your fragrance. In this case since we want to produce four gallons of fragrance, we can start by weighing a pint (16

Once you have your formula down, you can have your fragrance produced on a larger scale. The drums seen above are from projects of my own. Notice the red "Flammable" labels that warn of potential fire hazard. These labels are required as the contents of both drums are about 80 percent alcohol. The black drum on the left gave us about 800 2-ounce bottles of a men's cologne. The blue drum gaved us about 1500 1-ounce bottles of a men's cologner. In the case of the black drum, the fragrance was produced for us by a small fragrance house and then sent to a filling house to have the alcohol added. For the blue barrel, the alcohol was provided by the same firm that provided the fragrance. This barrel was a reorder of a fragrance that had sold out. The original supplier, a small company in New Jersey, had been purchased by a major French fragrance house and they provided this new supply for us.

fluid ounces) of the fragrance. There are two pints to a quart and four quarts to a gallon so once we have the weight of one pint of the fragrance we can multiply this weight by eight to get the weight of a gallon. Then, for this example, we multiply the weight of a gallon by four to get the weight of four gallons, which is what is required.

Once you have your formula, whether developed by weights or by volumes, you can produce whatever amount you need. When needed, the math for conversions is very simple. You are now ready to fill a few bottles or to produce your fragrance on an industrial scale.

The truth is that, when you're creating a new fragrance, you can never know in advance how others will react to it. Sometimes, like a new style of music or a new haircut, a new fragrance takes time to catch on – you never know. And that's why you always want to keep a record of what you've done. You want to develop a library of your formulas. Don't loose out on a good payday. Keep that notebook handy while you are creating!

Here are a handful of my own creations. Some were more successful than others but I have the formulas for all of them. As tastes change, some may be set aside; others brought to the front. If you are going to create fragrances, keep creating! And keep those formulas! You never know why or when they may suddenly be in great demand.

Appendix
Where to get stuff

You'll find my major list of sources in the "Vendors" pages of my PerfumeProjects.com website:

http://www.perfumeprojects.com/perfume/vendors/vendor-menu.shtml

Remember, companies are not around forever and, over the years, some of my favorite sources have disappeared. No doubt, in the years to come, more will vanish. But I'll be searching out new ones and they will be listed in those "Vendors" pages.

The important tools for developing production formulas are the electronic balance and glassware. Your sources are the same as those that would supply high school and college chemistry classes. You will find a number of these online. Some ship internationally at reasonable rates. As best I recall, my own balances came from "Affordable Scales and Balances" — http://www.affordablescales.com. They offer both "precision balances" (the category you want) and "analytical balances," balances with 4-digit (0.0001) readability, more sophisticated features, and higher price tags. The models I use are no longer available.

Made in United States
Orlando, FL
04 November 2026

90214008R00031